SHORT'S STORIES

BOMB DROPPING
NAME DROPPING
BALL DROPPING
MOOSE DROPPING
JAW DROPPING

Short, Ronan
Short's Stories: Bomb Dropping
Volume 1: Hampstead, North London 1940–1949
ISBN 978-0-578-59102-5

1. Biography & Autobiography / Memoir. 2. History / Europe / Great
Britain / 20th Century. 3. Science / History.

For more information, contact the author at ronanshort@hotmail.com

Printed in the U.S.A.
Distributed by Ingram

Cover: Interesting photo.... No, not the 'twerpy' kid in his new school
uniform—the building in the background, where Edward Mellanby,
later knighted by the King, discovered Vitamin D.

SHORT'S
STORIES

Volume 1
Bomb Dropping

Hampstead, North London
1940–1949

RONAN SHORT

Short's Stories

For

Keenan Charles
Gillian Klara
Jack Douglas
Rose Esperanza

Life can only be understood backwards;
but it must be lived forward.

—Søren Kierkegaard

BOMB DROPPING

The bomb missed…well, it didn't entirely miss. It glanced off a tall chimney of the Institute where we lived and exploded on an unfortunate house down Frognal, about a mile away.

The Hampstead area of North London is crowned by Hampstead Heath. The heath is five miles from the center of London and comprises about eight hundred acres of woodland, ponds, parkland, and common grounds, all former grazing areas. It sits at about five hundred feet elevation and commands great views over London.

The bomb in question was a German V1, a flying bomb powered by a primitive kerosene-powered jet engine—the world's first *cruise missile*. We called them *doodlebugs* or *buzz bombs*. We learned later from our Uncle Fred, a decorated London fireman, that as long as you could hear the motor running, you were safe, because when the engine cut out there was a fairly long glide path and then death, or at least destruction. Or a big hole that we could ride our improvised dirt bikes in. *Oops*, getting ahead of myself.

I was born in New End Hospital, Hampstead, North London, on March 6, 1940. My father, Douglas James Underwood Short, was in the pub opposite with the other expectant fathers, impregnating pints of bitter into their abdomens with shots of adrenaline. More about the *Underwood* bit later. My mother, Rose Margaret Mays Short, was in the hospital because, well she was needed there: I needed her there.

Life was busy. My older sister Gillian was four years older than I. She wore real clothes and seemed ready for school! I said *busy*. Oh there was all the usual things—feed and water and change my nappies. But Gill was learning real things and would soon be at New End Primary school. Another busy bit came from the air, courtesy of The German Luftwaffe.

Now Hampstead in 1940 was a scruffy, dirty, wheezing place with narrow alleys, soot-encrusted buildings, cars and lorries struggling up the hills, not to mention horses and carts still dragging heavy loads all over the area. The people, however, were remarkable: artists, writers, scientists, eccentrics, lords, and lay-a-bouts, and even Pearly Kings and Queens. We were all taught to wave at the German airplanes like it was a big joke while our parents *busied us* down into the basement of the Institute or, if there was little time and we were out on the streets, the Hampstead Underground station. The Tube station was safety—the real deal—four hundred feet below the surface, the deepest subway station in the world. Large wood and steel cages called *lifts* sent passengers slowly down on cables, or you could walk about six hundred steps in a long circle around the lift shaft. We all loved going down there and even as an infant I sensed the excitement.

On the platforms were camp cots, all laid out with blankets, and usually live music and lots of beer. Sometimes hundreds of people crammed the platforms and everyone partied merrily because you never knew when you came up in the morning who might be missing and whose house had been blown up or damaged. The power to the rails was shut off to keep everyone safe.

When we emerged in the grey daylight, people moved quickly to help the newly homeless or grieving families. It seemed that the Red Cross vans were always around with tea and buns and words of support.

We lived in a small flat at what was really government housing, no room to take people in, but the two almost unheard of things we had during the war was an endless supply of hot water and central heating! We had a very large bath tub—in the kitchen—with a board over it for a countertop. Many times two school chums, John Hunt and Terry Stone, from the large nearby working-class area, came over to play, looking dirty in scruffy clothes, and went home all cleaned up. Often I'd be told, *Don't go in there Ronan, there's three girls in the tub!* Or Gill and Sally would be warned of boys in there. We were known often as, *The Shorts—you know—the hot water people!*

The North London Consumption Hospital at Hampstead, built 1885.

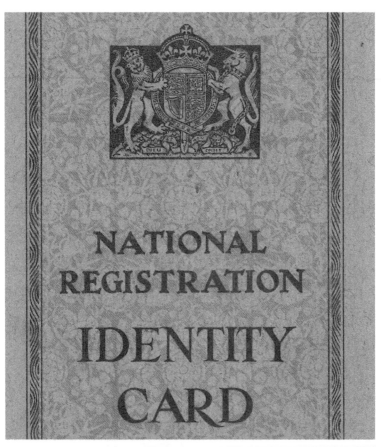

BOMB DROPPING

N.R.107A

NOTICE **DH** 228673

1. Always carry your Identity Card. You must produce it on demand by a Police Officer in uniform or member of H.M. Armed Forces in uniform on duty.

2. **You are responsible for this Card, and must not part with it to any other person.** You must report at once to the local National Registration Office if it is lost, destroyed, damaged or defaced.

3. If you find a lost Identity Card or have in your possession a Card not belonging to yourself or anyone in your charge you must hand it in at once at a Police Station or National Registration Office.

4. Any breach of these requirements is an offence punishable by a fine or imprisonment or both.

FOR AUTHORISED ENDORSEMENTS ONLY

CERTIFICATE OF OFFICIAL CAPACITY.

N.B.—No entry may be made here without express authority and after completion of the requirements as to " Identification Particulars " on the back.

I CERTIFY that the person to whom this Identity Card relates is employed in or under (is a member of) :—

Service or Department *Wardens Service*

Date *18·VII.43*

Signed *Paul Lawson*

Rank *District Warden* on behalf of :—

(HAMPSTEAD — WARDENS SERVICE stamp)

Service, Department or Authority.

National Registration Identity cards were carried by everyone except children under 14.

SHORT'S STORIES

The National Institute for Medical Research, acquired in 1918. Our flat was next to the tower on the right.

Borough of Hampstead

AIR RAID PRECAUTIONS

This is to certify that D. J. Short Esq.

Holly Hill N.W.3

of National Institute for Medical Research has been duly appointed as an Air Raid Warden. This is his authority to carry out the duties laid upon wardens by the Borough Council.

P. Marwood
Town Clerk.

Date of issue of Card 8 - FEB 1939

Date of appointment of Warden 26/9/38

Signature of Warden Douglas J. Short

TOWN HALL,
HAMPSTEAD, N.W.3.

My dad was appointed an air raid warden on September 26, 1938—a year before the war began—and was active until the end.

How to Catch a Wife

Most of the time when I was very young we stayed in and around the small flat at the top of the Institute where we lived. That is the National Institute for Medical Research, Holly Hill. It was a former tuberculosis hospital named Mount Vernon, purchased in about 1918 by the Medical Research Council. If an air raid was called and there was no time to walk down the hill to the Tube, we went down into the basement of the building.

Dad's day job was as the senior animal technician, and he became an Air Raid Warden in the evenings. As an Air Raid Warden, he would report to Post 15. They would wear blue "siren suits" (heavy coveralls), carry gas masks, and wear steel helmets with the letters *A.R.P.* (Air Raid Precautions) on the front. There were about a dozen people in his post, including various notables such as Jill Craigie, Judge Broderick, John Gaudioz the registrar, Ivor Rees-Roberts, and others. All were considered in essential jobs and excused from active duty. Now Jill Craigie, a model and professional photographer, was the mother of Judy, and later the wife of Michael Foot the leader of the Labour Party. Judy was and still is Gill's best friend. As a small boy I would regularly mess up all their day's plans when my mother would say, *Take Ronan with you!* These words would be uttered later to constrict my own adventures...*Ronan, take Sally with you!*

The ARP wardens, as they were known, would walk certain *beats* and during an air raid warning, *a blackout*, they would order lights to be put out, curtains drawn or stop cars that had headlights on. They would also go up to a couple of anti-aircraft gun emplacements and deliver important messages—and of course tea and biscuits! When it was quiet they could sleep at the post and go straight to work in the morning.

Rose Mays Short, about the age of 20, with the latest hairstyle, the *Marcel Wave*.

Dad was born in December 1904 in Acton, West London. His father was an auctioneer in London but died young at age 63 in 1926. My mother was born in Watford, in August 1905. Her father, Frederick Mays, was an engineer/millwright at a large commercial bakery run by the Seventh Day Adventist Church.

They were not practicing Adventists, but life in the Mays household was strict. After high school in Watford, Rose was trained as a *piper*, a person who made expert and creative designs on cakes, confections, and chocolates. She had steady work and spent many a weekend playing tennis with friends at the West Herts Cricket and Tennis Club, where she met my father. Doug had arrived with some teammates to play cricket one Saturday when he noticed this young attractive lady playing very good tennis.

Dad, seated center, was an early captain of West Herts Cricket Club in Watford where Rose was raised.

"Who's that?" he asked.

"Oh that's the Mays girl," replied one of his mates.

"I see," said Dad, smitten.

He stood, transfixed. It was a *love match*!

"Well that's the girl for me."

"Don't be silly Doug, there'll be plenty more at the dance tonight," said young Alan Grimsdell.

"No, I'll see her later in the pavilion. She's the one."

And she was. They quickly started *walking out* and were married in May 1932 at the Hampstead Register office. Both of their fathers had passed away by the time of their wedding.

Doug moved her from the restricting lifestyle of the family home in Watford to the Institute, where he was already living. Gillian Mary was born on July 6th 1936, and I followed in 1940, just in time for the war, and Sally Margaret in October 1942.

Not their wedding—which was an ordinary affair at a register office—but attired for a much classier event.

Flying Monkeys

At the top of one of the hills in North London is Hampstead Heath, a semi-wild expanse of ponds, woods, fields, and groves of birch trees—in hundreds of acres. At the very top is the Whitestone Pond, an ancient pond fed by an artesian spring with shallow ramps at each end for horses and carts to enter and exit. No fish, but a great place for sailing toy boats and for dogs and other animals to drink.

The donkeys that quenched their thirsts came from the donkey rides that hung out every summer on the heath. They were tied up near the enormous cast-iron drinking fountain with chained-up solid brass cups. For a lot of the war, the pond was drained—it was considered a highly reflective bombing target. Just adjacent to the fountain was a tall white flagpole, rather formal. At the top of that flagpole sat the second escaped monkey. Dad fired both barrels! The story begins like this....

The Medical Research Institute had literally thousands of animals in its care. Most of them were small mammals: rats, mice, and guinea pigs. But it also had larger animals like dogs, cats, and monkeys and a few exotics like armadillos, marmosets, and a couple of chimpanzees. The chimps, named Swan and Edgar, after a posh London store, occasionally had dinner with us and they played with their food to such a degree that we were scolded for laughing at them.

The monkeys, *rhesus macaques*, were used in various studies to find a cure for polio, which was a scourge in Europe and the US. On this particular day, a new employee left a large cage area open for a second as he reached back for more food and about six of them were gone!

Four of them did not go far. They got down into the grounds and onto the tennis courts where they defecated and then threw poop at each other over the net. No game there. They seemed set, but the racket was stopped with bananas!

The extreme leader went farther, over the ivy-covered wall and across the street to a large private house with an official flagpole in the front garden. Dad saw it from his office window—it was happily sitting at the top of the flagpole—and he acted quickly. Grabbing his 12 gauge shotgun from its locker, he rushed down and across the street, straight onto the immaculate front lawn. Without further ado, he proceeded to shoot the monkey off the top. The limp body fell at his feet. He stuffed it into a burlap bag and sped off just as the homeowner came out yelling loudly. *Oh, for a silent shotgun*, Dad thought as he ran back to his office.

The homeowner was not silent. He was in fact The Right Honourable Ramsay Macdonald, a former Prime Minister of Great Britain.

Before he could get back, one of his technicians ran up to him shouting, *Sir, there's another one on the flagpole up by the pond.* An expert by now at shooting errant apes from flagpoles, Dad ran the few hundred yards to the Heath and in front of several startled strollers enjoying the vistas and the peace and quiet, dispatched the second one.

He was barely back in his office when the phone rang. It was Sir Henry Hallett Dale, the Director of the Institute demanding my father to appear in his office *NOW!* 'Sir Rinry,' as he was known by all, was no mean slouch himself. He had been in physiology research most of his life and had been awarded the Nobel Prize in Physiology/Medicine in 1936.

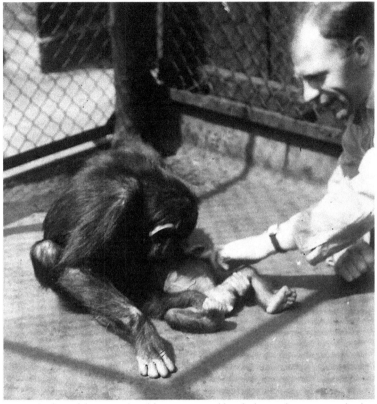

Doug playing with Swan the chimpanzee and her baby at the MRC. He had
an innate connection with lots of animals, especially primates.

"Short, you are a disgrace to the Institute," he began in a
very loud voice.

My father, slouching over responded, "Yes sir, but I
was only trying to save the Institute from an embarrassing
situation."

"You," replied Sir Henry, "are an embarrassing situation!"

"You *should* be fired and I *may* be!" he stormed. "Absolutely
outrageous, and worse still—on the former prime minister's
front lawn!"

The harangue went on for twenty minutes.

"You shot a monkey from the private flagpole of the
second most powerful man in Britain, then you ran away and

did the same thing up on the Heath in a public area ten minutes later! They were important experimental animals in the polio study."

"Yes, Sir," said Dad, wondering where he could get another job. "On the other hand, sir, those research monkeys might have got completely away and the press found out."

The boss was unconvinced. "I have a good mind to give you the sack here and now!"

Dad was now silent. He'd never seen his boss this angry and his knees were shaking.

"Get out," said Sir Henry.

Dad turned and stumbled towards the door, grasping at the handle for support.

As he opened the door Sir Henry, in a much lower, almost normal voice said, "Short!"

"Yes sir," said Dad, turning around hesitantly.

"Well done, Short. Well done!"

How to Catch a German Spy

As I was told, one night I apparently became restless and Rose took me out of the only bedroom away from my sleeping father.

She nursed me in a window seat in the living room that overlooked streets and houses from its 6th-floor position. As she looked out, she noticed a flashlight dancing across an expansive flat roof, lighting up what looked like a large metal clothes dryer.

That's odd, she said to herself, *It's a blackout—no lights allowed.* The light lingered a while and then disappeared. "I must tell Doug in the morning."

The next day she tried to tell him but he was too busy. *No time now, tell me later.* he said as he hurried out the door.

The next night, as babies do, I awoke again at about the same time. So there was Rose sitting in the window when the light again moved slowly across the roof.

"Doug," she said in the morning, "I have to talk to you."

"Rose, I have a meeting with the director—police are here, talking to all the foreign scientists!—I can't say anymore, must go."

Well, it was couple of nights later when I next disturbed the family bedroom and sure enough the *night light* appeared again. The following morning she got up earlier than usual, took my Dad a cup of tea in bed and poured out the story.

"Good Lord, Rose," he sputtered. "That's what this flap is all about. Someone—believed to be a German spy—is transmitting Morse code and/or radio messages from right here, or nearby. Everybody is on high alert. Why, woman, didn't you tell me about this before?"

"Oh Doug, I tried, but you were too busy to talk!"

The next night two policemen were on watch in our living room. I had done my job so I slept through the night like a... well you know! Soldiers and armed police arrested the man the next evening.

The story that I grew up with was that he was summarily tried, convicted, and was one of the last to be executed by firing squad in the Tower of London. The low point of his life turned out to be the high point of mine, and I wasn't yet a year old!

Rose holding Ronan as an infant.

HOW TO CATCH A GERMAN SPY

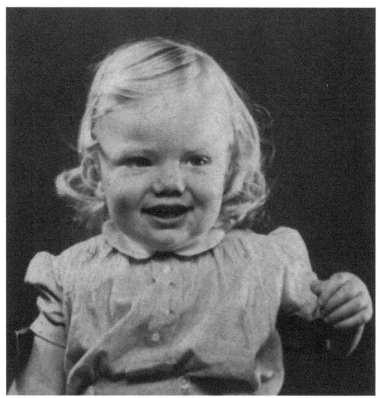

Young Ronan with long curly hair and likely a hand-me-down dress from Gillian!

From our window high in the Institute, Rose looked down on rooftops as she nursed me.

Evacuating Away

One of the first things that happened to us kids in Hampstead during the war was…leaving. We were all evacuated, when old enough, with our gas masks and name tags to different parts of England.

Sally, my younger sister, went to Uncle Alf and Auntie 'D' Knight in Watford. He was also a freemason and owned a large community bakery and delivery vans. There as a tot she helped in the shop, had flour put on her nose daily by the shift bakers, and ate a lot of sticky buns. She also got to ride the *speechless server*. Up and down pulled by cables, it transported all the bread and cakes from the ovens in the basement to the store, aka *the dumb waiter*.

Gill went first to Auntie Sybil in Northampton and her small, dark cold house. Life with Sybil was sparse, to say the least. Sybil was a struggling street trader in local markets where she sold exotic fabrics and re-packaged saccharine tablets! Gill was the re-packager. It was her job while she played with fabrics in the chilly house to lay out tablets from a large sack on the floor onto a special tray contraption, which then counted and delivered 100 tablets into pre-printed envelopes. She licked the flaps and sealed them, ingesting perhaps enough *stuff* to sweeten a cold, or prevent cholera, which was excellent

Sally, Ronan, and Gillian. This posed arrangement was repeated at least three times as we all grew older.

because deadly cholera had actually started to appear in the city water supply.

Dad heard about this, came quickly care of the *Red Cross box*, and took her to Birmingham to Uncle Sid and Auntie Margery's house. Auntie Sybil had lived a difficult life from early years. She was the third youngest of ten children and—although she showed great promise as a young dancer—was actually given away to another family to raise. The circumstances around this event can only be imagined. Sybil was naturally very resentful, and she reflected on it all her life.

Many years later, my wife and I and our two daughters, then about ten and twelve, visited her at her council flat in Northampton. She was quite elderly and a little unsteady, but we had a lot of laughs at lunch at a nearby hotel. We went back to her flat and she wanted to know about Alaska. *Alaska*, she said. *Most of my life I couldn't even afford to go to London!* Her flat was warm and cozy and nicely furnished. She loved the fact—and so did I—that her flat backed right onto the Northampton Rugby Club.

She served tea and scones with jam and then disappeared for a minute. When she returned, she was quite emotional.

"I want you to have this," she said, pressing a coin into my hand.

I did not look at it right then, as she was talking:

"I don't have much, Ronan, but your father was my older brother and he helped me out a lot during the bad times."

I opened my hand and saw that it was a US Liberty gold coin. I treasure it, and even now it sits in a safe deposit box.

My evacuation journey north was taken as a *Red Cross transferee* to join Gill at Sid and Marge's house on Fountain Road, Edgbaston. Birmingham was, and still is, a large industrial city in the midlands. We had, however, to wait weeks to go the hundred miles or so up north, as gas was heavily rationed—unless on official business. Dad got fed up with waiting and one day glimpsed in the back of a closet at work a big metal trunk with large *Red Cross* insignias on all sides. *Just the thing*, he must have thought. He strapped it on the luggage rack at the back of our Ford 8 Saloon car, licence # *DXR239*. He then obtained gas easily, *On official business, transferring refugee children*, he would say pointing to the black trunk, *Oh, of course sir*, said the gas station attendant, filling the tank. Well, we were children-yes, but refugees—no!

The Red Cross had moved me from a moderately bombed area to a heavily bombed one! However, they had a nice house that turned out to belong to Sid's sister, and Uncle Sid had a job—which transpired later to be quite a rarity. Gill's recollections are the house had little furniture and no toys. They had no children of their own and little parenting skills. They gave us an empty room on the second floor where we could make papier mache *things* with newspaper, water, and flour-paste glue. In the summer we played in the garden and an old summer house with lots of water and a black cast-iron pot. We mashed up all the hollyhocks in the garden and made *witches potions* with which we dyed everything, including ourselves, bright purple.

The *black box* ruse not only worked to take us to our relatives, but later for our parents to visit us. *Oh yes, known official vehicle*, would be heard at the gas station, *no coupons needed*.

Growing Up in Hampstead

My younger sister Sally was born in 1942 and by 1948 was, in my eyes, a real human being—not a screaming, squawking, crawling thing, but ready for fun, like running around on Hampstead Heath, falling in the swimming ponds, teasing the wallabies in Golders Hill Park, and other enriching activities.

The closest part of the heath was about 250 yards from the Institute. We would usually head for a glade of silver birch trees where we made *camp*: a few sticks with our coats over them and we were in darkest Africa, or darkest Essex. We had climbed all the birches and felt comfortable doing it. That is, until the day I slipped and fell, hitting several branches. When I was about fifteen feet from the ground, the braces holding up my short pants caught on a jagged branch sticking out of the trunk. Sally tried, but couldn't get me off the hook! So she ran for home and I started yelling. I was up there for about two years—well, it seemed that long—when some walkers heard me and after a bit of a struggle got me down. I had a bloody nose, torn shirt, and various scratches and bruises when my father arrived.

"Well, son" he said checking me out, "thank your rescuers properly!"

Which, of course I did.

As we walked away he berated me for interrupting his tennis match. "Fancy disturbing a couple's quiet walk on the heath. Sally gets full marks and an ice cream for running straight home, but it's straight into a bath for you."

More hot water!

A lot of the time we would play in the streets with other school kids and during school holidays for lunch we would often go to the Moreland Hall, a sort of social center that had a large kitchen, a stage, and chairs and tables. There was food there for old folks, handicapped kids, refugees from Eastern Europe and *street urchin gangs*, aka us. Sometimes for sheer sport we would jeer at or taunt the foreigners we saw on the streets for having long beards and strange clothes and even stranger speech. That is until getting home one day I was abruptly confronted by our parents! *What is this I hear?* You, *a member of our family, Ronan, being rude to new arrivals, people who have greatly suffered in the war and that Britain has given refuge to.*

Smack! against my left ear, and then *Ouch*, my mother hitting me with a thin cane across the backs of my knees. Then sent straight to bed with a bowl of bread and water, and maybe a Lincoln biscuit later when they'd calmed down.

It didn't end there either. We were soon taken to see these refugees at the Moreland Hall and forced to profusely apologize. In time, we were tolerated and they began to wave and say hello, and we would even help carry their groceries home from Sainsbury's. Some of the young Jewish boys started to go to the heath with us. Other boys would shout *Yids! Yids!* and we would reply, *Yes, but they're our yids so shut up—leave them alone!*

We would always be given free fruit at Nethercott's, the greengrocers on Flask Walk. Sid was a Freemason like Dad. Next door—if we had a farthing—we bought a glass of *Tizer* "the Appetizer," a kind of cola that was made right there in an elaborate machine. Flask Walk was a tiny, narrow street about a block long. We would go in the Flask pub next door and ask

The Flask Pub, established on Flask Walk as early as 1663, still stands and is a famous landmark known for its mineral water. We only ever entered the pub to ask for a drink of the water.

for a drink of water, which if you were lucky, came with a drop or two of orange squash added. I was now becoming an eight-year-old delinquent, but this was slowed by my mother's eternal instruction: *Take Sally with you!* This slowed me down, but speeded her up, as she was forced to climb and run as fast as the boys. And run she did, despite being a short-arse, but I'll get to that later.

Another time I was in the Tizer shop and a couple came in right behind me. They were old, at least fourteen, and they waited as I ordered. She had piercing blue, almost purple, eyes and dark hair. I was mesmerized and couldn't stop looking.

When the assistant said, "That's a ha'penny, please," I could only fumble in my pockets.

"I'll get it," said the young man with a foreign accent—American, or something odd.

"Thanks a lot, mister," I said, still staring at her.

SHORT'S STORIES

I sat down to drink my Tizer, in a glass of course, no wandering off, and I just wanted to stare anyway. They each got something, looked around for a minute, and left.

"That's Libby Thomas," said the shop assistant, thinking. "She's from Wildwood Way, over by the Spaniards Inn. I think she's in a movie. Uh...*Liz* Thomas."

I finished my drink and was exiting the shop when she said, "No, wait, it's *Taylor*, that's it, Taylor."

Cubby Hole: No Toilet

Our flat at the Institute was originally a one bedroom place for a single person who would be in charge of the animals and on-call (i.e., Dad). The bathtub as I have said was in the kitchen with a large board over it for a countertop. In the one bedroom was a tiny narrow closet with a sink at the end and a wooden step: the cubby hole. But the flat had no toilet. That's right. *No Toilet!*

On our floor at the top of the Institute was a long corridor. To the right of our front door was a large room with a brown linoleum floor that held lounge chairs, dart boards, table tennis, and a full-size snooker table. Any misspent youth I might have had began there and all our large holiday and birthday parties were held in the *snooker room*, with boards placed over the exquisite and well-kept green baize.

There was another flat next door to the left where Jack and Vi-Vi Cleve lived--with matching-no toilet cubbies. Just down the hall on the left was the elevator—a Waygood-Otis—and just past it was, yes, a toilet. In fact two, with floor-to-ceiling Victorian partitions! And farther on down were two more. These two were opposite the canteen, where mid-morning coffee, hot lunches, and afternoon tea were served in separate rooms for the scientific and technical staffs. Sometimes all toilets were busy; it was like waiting at a train station to use the john. The good news was that we never had to clean a toilet, it was all done by the institute cleaners. When we moved to our flat in Mill Hill I asked when the cleaners were coming in. *They're not*, said Dad, *so get cracking son! The stuff's under the sink!*

Oh, the horrors of knowledge and labour.

LIFE AT THE INSTITUTE

The National Institute for Medical Research was a world famous place. There were about three hundred scientists and technical staff that worked there every day, with several Nobel Prize winners and many more to follow. Sir Howard Florey did early research on penicillin with Alexander Fleming, and Sir Henry Dale on neural transmitters. Oxygen deprivation studies had been done there for many years and all the altitude sickness testing for the 1953 Everest Expedition was undertaken there, led by Dr. Griffith Pugh, the physiologist on the successful assault.

Sir Charles Harington was made the director in 1945, following Sir Henry Dale. He and his wife and family lived in a large house on the grounds.

Although most people left around 5:00 p.m. every night, you could find scientists and staff working late or all night. The two I remember best were maintenance workers: Albert Onions and Tom Waring. Albert was the night stoker, keeping the boilers full with coal and later, coke. He came back from the war with shell-shock, now known as PTSD. A large man, he was quiet at first, but as time went on, I believe he healed himself in two ways: having a steady job and whistling, *yes, whistling!* He gradually developed the most beautiful sounds— tunes, birds, classical themes, and more, all the while working outside shoveling tons of coke into the furnaces. His sounds echoed off the tall brick buildings and we could hear him many evenings from up in our flat.

The other chap was old Tom Waring. He was the night watchman. On his breaks he would descend into his "office" in the cellar of one of the corner towers, named *Dew Drop Inn* and—if Gill, Sally, and I were around—he would invite us down too. It was a tricky, vertical steel ladder. Gill and Sally would occasionally sit on his knee and twiddle his large handlebar moustache while he told stories of the war. After a while, we got what we came for. He would save up empty cans of sweetened condensed milk, the lids of which were still partially attached with sharp jagged edges and razor-like points sticking up with tantalizing drops of pure goodness attached to them. He only had a rusty old pocket knife to open them. We would cut our tongues, our fingers, and our lips trying to get every last bit of that delicious stuff.

Whenever I see that Eagle Brand condensed milk today, I have flashbacks: my mouth waters and I almost feel for the cuts!

This is the can without the jagged edges.

Visit to the Godmother

Life as a child divided itself between bomb-touching on the one side and sublime visits with my mother to houses like Coca's, Judge Broderick's model railway, and my godmother: Hilary Rees-Roberts. Hilary lived farther down Frognal, the road next to the Institute that was named for a long-gone frog pond. Hilary lived in an elegant Edwardian house with a double front door and elaborate stained-glass work everywhere you looked.

I usually enjoyed going there, but this day was different. Mum and I started out strolling along in the spring sunshine. We had gone a short distance when she started in at me.

"You are a big disappointment, Ronan. You stay out too late after school. Your homework comes back heavily marked up and that's when you haven't lost it, and you actually do it.

"But Mum, it wasn't my fault that I kept banging the bass drum. The other boys told me to."

"Listen to the teacher, not the other boys." *Now, there's a new concept*, I thought. "And last week when you didn't come home till nearly 9 p.m., your father was down at the police station and I was ironing the laundry. I didn't need the new-fangled steam button, water was dripping down nicely and keeping it all damp. I did everything three times." She wasn't finished. "You are the worst son anybody could have. You were supposed to water everything in the garden. Dad said you haven't been anywhere near there by the looks of it. You know the tomatoes need extra, now they're dry and wilting."

By the time we got to the elegant front doors I was down in the dumps and ready to turn around and go—anywhere. The doorbell was answered by her maid Bernadette.

"Please come in. Miss Hilary will join you in a few minutes. I'll show you to the drawing room."

The hallway had framed paintings of long dead relatives. We sat ourselves down on elegantly upholstered chairs and waited. Neither of us made eye contact or spoke.

Eventually Hilary entered. "Rose, lovely to see you again, and Ronan, how are you?"

"I am well, Mrs. Rees-Roberts, thank you," I stammered out wishing for fresh air, any air.

"How is Ronan doing at school Rose?" said Hilary turning away, slightly and excluding me.

"Doing?" said Mum, "He is doing very well indeed!"

I misheard her and my jaw dropped towards the floor.

"Oh yes, he has fine marks in all his subjects and he's singing in the church choir, soprano soloist occasionally, and vice captain of the choir cricket team. They played against Christ Church last week and Ronan scored the winning run and *carried his bat*. He plays football too for New End School. He is the apple of my eye and of course Doug is so proud of all he's accomplishing he's *almost* speechless."

I *was* in shock and speechless, but managed to scarf down the elegant tea sandwiches, *petit fours*, and chocolate cake that Bernadette had set down in front of us. After a lot of chitchat that flowed above my head we made our farewells, departed and walked up the hill towards home.

At one point, Mum bent down to brush the hair out of my eyes and I saw her wet cheek. I felt so confused and so honored I could have kissed it—so I did.

"Thanks, Mum," I blurted out.

We both stood for a long moment holding hands and smiling at each other.

P

LONDON COUNTY COUNCIL
NEW END J.M. & I. SCHOOL

REPORT FOR YEAR ENDING *July*, 1949.

Name *Ronan Short*   Number in Class **44**

Class *II*   Position in Class **35**

SUBJECT		ASSESSMENT	REMARKS
ENGLISH	Oral	C+	Good, clear speech.
	Written	C+	Above average.
	Reading	A-	Very good reader.
	Comprehension	C+	Intelligent.
ARITHMETIC		C-	Weak - needs more effort at home to ...
HISTORY		C+	Interested.
GEOGRAPHY		C	Average
SCIENCE		C+	Interested in N. Study.
ART		C-	Rather poor.
WRITING		C	Average mark.
HANDWORK		C-	Rather lazy.
OTHER SUBJECTS	P.T. Games	B	Keen on subject
	Swimming		

RELIGIOUS KNOWLEDGE *Good*   Attendance *Excellent*.

GENERAL REPORT *Ronan missed his English exam. This has made him so low on the class list. His conduct leaves much to be desired.*

Class Master   *A.D.D.*   Head Master

My report card. I agree with the intelligent part but question the rest.

Chicken with French Dressing

Just beyond the aforementioned tennis court and against another wall my father kept chickens. Lots of people had a chicken run during the war. I was now about four. They also had *allotments*: small communal gardens to grow fruits and vegetables. We would feed our chickens table scraps and I would add floor sweepings from the Institute grain store with an occasional scoop of *borrowed* good stuff from the large sacks staged along the walls.

Across the street from the Institute grounds was a row of large houses. Former Prime Minister Ramsay McDonald lived in one and a certain General Charles De Gaulle in another. De Gaulle had been evacuated by SAS commandos after the fall of France in 1940 and was now head of the Free French Forces in Britain. A very tall man with a deep commanding voice, he would later become the president of France.

One day when we were feeding the *chooks*, loud knocks banged against the small wooden door in the brick wall.

Dad opened the door immediately. "This is a private side entrance, you must check in at the front gate!"

Two large men in grey suits and even greyer accents said together, "You have a visitor".

CHICKEN WITH FRENCH DRESSING

General Charles DeGaulle lived a few streets away in a house provided during the war by the British Government.

"Visitors?" said Dad. "Front gate," and started to close the door.

A deep, accented voice came from behind the men. "Please allow me to introduce myself...I'm Charles De Gaulle."

Dad looked at the very tall imperious figure and said, "Well, Blimey, you might be! I'm Douglas James Short. My son—Ronan."

"I am he , how do you do!" he smiled, extending his hand to Dad. And then bending his enormous frame down to me, said, "Please tell me about your chickens—I have chickens too!"

His two aides, aka bodyguards, drifted into the background.

"How do you keep your chickens so healthy and active?" he asked.

"Oh," I piped up, "we feed them government gr—".

Dad interrupted with a look that could kill a chicken. "Er—scraps—food scraps. Don't they look good?"

I was shooed away to put more feed… er, scraps, down in the enclosure. De Gaulle went on to express his great gratitude to British commandos for rescuing him at the Fall of France and bringing he and his entourage to Britain. They continued to adult talk until, finished with feeding, I moved back to the group.

As the conversation was coming to a close I repeated Judy's now famous question. "So, General De Gaulle, can you tell me, sir? Do your chickens cluck in French?"

ELEMENTARY SCHOOL
AND THE UXB GANG

The war finally ended with a bang—two actually, in Japan—but it had ended in June, 1945 in Europe. There were parades and military bands, many ceremonies, and flags being sent up monkey-less flagpoles. All this was followed by the voting out of office of Winston Churchill. At school (I was about five) we were taken to the streets and cheered all the celebrations—they went on for months. Pubs paid off their mortgages. We were sometimes escorted by grown-ups, as the streets were pronounced *very dodgy* by our teachers: full of lots of soldiers who had been *over-served*, as our nervous headmistress put it, and liable to commit *foul deeds*.

This was quite untrue, generally, as we would find out. After school, on a regular basis several of us would go to Dick's, aka Charlie's the tuck shop, and buy lemonade powder or penny gobstoppers and then go straight home. *Er...no, that never happened*. We would play soccer in the streets with an old tennis ball. And on one particular occasion we were quite a long way from our *patch* and came across a bombed out house surrounded by nailed up corrugated iron sheets and big UXB letters painted on the side. The oldest boy in our group was Johnny, a Jamaican kid afraid of nothing so we all looked up to him.

"There's a bomb in there," he said. "It's still there in the basement."

"How do you know?" the other three of us said.

"Well, the house was destroyed by the bomb falling, but several houses would be gone if the bomb had actually exploded, hence the *UXB* on the fence. Let's go in."

He pulled away a corner of a rusting metal sheet.

"Wait!" we said.

"No, it's OK, the UXB sign is for old ladies and the public," he said, "not us!"

We did this on several houses. We would climb down gingerly through the filth, the jagged timbers, the broken furniture, smashed doors, and wrecked gas stoves. We pushed everything aside until the bomb was visible in the destroyed basement area. We would take it in turns to reach down and touch the body or the fins, whichever was closest. On this particular day, it was my turn and I would be King of the Castle if I touched it. Just as I reached out and placed my hand on the fins a loud hissing sound and a slowly increasing ticking noise came from inside! *Really? No! Just making sure you're still reading!* No, I was declared King for the day and then we climbed out, filthy dirty, and made our way home. The bomb-touchers had struck again!

My way home was all uphill with lots of steps and several narrow, usually rain-soaked, alleyways. Lots of people about: noisy drinkers inside and out of the three pubs I had to pass, men and women coming up Heath Street from the Tube. Some greeted or nodded at me, some would know me and say hello. There were lots of servicemen and nursing staff from the two nearby hospitals. If we saw American soldiers, the standard approach was, *Hi, got any gum chum?* and the response was usually fruitful—*Juicy Fruitful!* There was rarely, if ever a sense of threat or danger from anyone on the street. We were now all together, as our enemy was Germany.

New End Primary School had classes for Infants to Seniors (3-11 years old). I attended until I was about 9, when we moved.

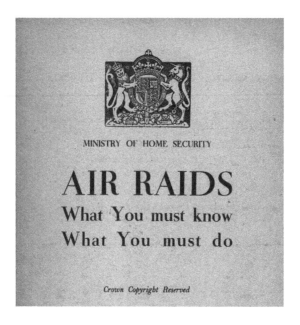

Every home had instructions for air raids, the most important being black-out curtains and directions to the nearest shelter.

Learning the Piano:
A Lesson

When I was six or seven, it was decided that, due to my general street urchin *m.o.*, I should take piano lessons. One day, a middle-aged woman appeared at our front door and I was instructed to take her coat and hang it up.

"Ronan, this is Madam Doreena," said Mum.

We shook hands and I immediately noticed she had furs hanging around her neck, and the heads of the former weasels, or whatever they were, were still attached and linked by a gold chain through their noses.

We were left alone and she started teaching me, in a rather insistent manner, to read music and play basic scales. This went on for several weeks—once a week I had to greet the two *weasel brothers*. The lessons got harder and I got further behind due to not practicing and studying the music theory part. After each lesson, I had to fetch Madam Doreena's outer coat, escort her down in the lift, and then formally thank her at the main Institute door.

I was young as I said, and completely ignorant of learning music, etc. After the lesson, I would usually go and lie on my bed and hide, as my hands and fingers were starting to hurt. I tried stretching my fingers as far as I could, but it still hurt and slowly got worse. *That's music lessons,* I said to myself—they're hard. By now I was hiding as soon as she left, as my knuckles were bleeding sometimes and staining the bedclothes. *Why am I so inept*, I thought.

Soon after, there was a concert at New End School and Joan Bottom (the first girl I ever kissed, and with a name like that I was teased mercilessly), played a very nice piano piece along with several other students on various instruments. Afterwards, they were all smiling and their hands looked fine and Joan's were quite tiny and soft.

"How are you holding up?" I asked her.

"Holding up?" she replied quizzically. "It's piano lessons, not cricket and football like you're playing all the time."

I was mystified...and I had a lesson coming up that night.

Miss Doreena arrived and we started, the weasel brothers gloating at me from their beady eyes.

We were about halfway through the lesson when Rose suddenly opened the door. "Excuse me," she said.

A tiny spurt of blood had just left one of my knuckles and landed on a white key.

"Failure must be met immediately with harsh treatment," said Miss Doreena, her ruler raised.

"Failure!" yelled Mum, grabbing her by the weasels, whose eyes now seemed quite large. "How dare you hit my son like that?"

She grabbed the metal-edged ruler and threw it across the living room.

"I shall report you to the piano teachers guild and I will notify my husband now and he may well call the police."

Miss Doreena, *the former piano teacher*, grabbed her own coat and fled down the many stairs and out into the street.

"How long has this been going on Ronan?" she said, cradling my hands in a soft, wet, washcloth.

"Almost since the beginning, Mum. I didn't know how music was taught."

Next day I told Joan Bottom. She held my hands for a second, then kissed me on the cheek. It was all worth it, even with more teasing.

THE UNDERWOOD CONUNDRUM

My father started at the Institute in about 1918 when he was fourteen years old. He was a volunteer in human physiology—a junior assistant. He received no pay, but free meals in the canteen and when they worked late he would be taken across the street to the Holly Bush Pub for a shandy—half beer and half lemonade. On those days he would go back and sleep on a cot in the exercise lab rest area instead of cycling many miles back to Watford. The testing they were doing then was for mountaineering and involved oxygen deprivation and exertion. They also were studying brown and black lung mining diseases.

One day his boss called him in and said, *We have enough money now to hire you Short, so you must get signed up in the office. You'll need your birth certificate!*

Dad went home and told his mother.

"Oh, I don't have anything like that—you'll have to go to Somerset House."

In those days all the records for England and Wales were kept there. So a few days later Dad is up in London at Somerset House.

"20th Dec 1904, Acton, West London. Douglas James Short? No, nothing!" said the registrar.

41

The family name is still unresolved, but my father was definitely born an Underwood. The story continues.

"But you must have," said Dad, "that's when I was born!"

"No *Short*," said the registrar. Then, after looking through several more records, he exclaimed, "But I do have a Douglas James Underwood born on that day to the same woman you describe and in Acton, West London!"

Thoroughly confused and afraid he wouldn't have a job, he paid for a copy of the Underwood certificate, went home, and showed his mother.

She didn't bat an eyelid, didn't skip a beat. "Oh really, well use that one!"

"But that's not me, Mum!" he exclaimed.

"I don't know, Doug, I can't remember, and I don't want to discuss it anymore!"

There has been an on-going question about how and why Underwood became Short. We tried asking our mother, father, aunts, and uncles, but all were as quiet as a Victorian living room. The day after she died in Croxley Green, near Watford, several of my aunts went quickly to a stout metal box under her bed that was known to be full of family papers. To their shock it was empty. Over the last weeks and months she had fed all the documents into a small woodstove in her bedroom.

A few years ago, I tried to find my granny's mother via a genealogical search, but to no avail. We have discussed many possibilities—that my grandmother was married several times, perhaps without divorcing the previous husband. Or that poverty and/or bankruptcy was behind a *midnight flit* wherein everyone ups and moves with an accompanying change of name.

TWO-WHEELED TRANSPORTATION

One day a large lorry appeared in the driveway of the Institute, a common occurrence, but Dad insisted I help unload. And there it was—a worn, used, dull black, scratched-up, two-wheel bicycle! Truly *a poor thing, but mine own*. I loved it.

My Auntie Nin, a governess, and the Tress family in Scotland that she worked for, had sent down their son Robin's old bike as he had received a brand new one. It had a torn sprung saddle, stiff solid-lever brakes, a rusting chain, and clumsy broken mudguards. *Just fine*.

Getting on and riding it was another matter. After a lot of bloody knees and gravel-embedded fingers, I was pushed off yet again. I turned back to ask Dad a question about which brake to use when, and like kids everywhere, I had a pivotal moment of my existence on the planet. He was way back and I was riding, pedaling, balancing, only to fall off, of course, once I realized it.

The old bike worked well for months, until one day Richard Price from round the corner near the church at 3, Benhams Place came over on his bike. *Oh my God*, it had twelve speeds, chrome wheels, shiny paint, cable brakes, drop handlebars, a racing saddle, and a leather saddle bag for tools and lunch! Everything had a brand name on it including lunch!

It was years before I rose to his level, but Richard taught me all about bicycles and how to fix them. I'm now doing it for my grandchildren.

Martindale Sidwell
and the Church Choir

My street urchin behavior was further restricted by a statement my father made, apparently after yet another act of errant behavior. *Son, we've spoken to the Reverend Carnegie at the parish church and you are joining the choir.*

The choir! I couldn't believe it. I was a street kid—part of an important bomb investigation team, no less. I was horrified at the thought, but a week later at the appointed time I was sent down the hill to the church and knocked on the door of the vestry at the side. A loud, rough voice said, "Come in!"

I entered to a cluttered room filled with papers, books, and piles of music scores under which was a grand piano. The figure seated there did not look up, but just barked. "Don't just stand there boy, sing something."

I was in the presence of one Martindale Sidwell. "I don't know what to sing, sir," I said feebly, "I'm new." *A weak attempt at Oliver Twist!*

He handed me some sheet music with a hymn that I vaguely knew the tune to, so I warbled away for a couple of minutes. His reply was swift, and though I awaited the inevitable and hoped-for, *Get out and don't come back*, instead he said rather softly, "Choir practice Thursday evenings, two services on Sunday—sixpence a week. Weddings are extra money, maybe a shilling. Negotiate with the bridal couple."

I was stunned. I was a professional choir boy—being paid money! What did *negotiate* mean? The first thing I had to go through was not negotiation, but initiation. A few days later I was taken by a senior choir boy down steps to a dark, dank doorway that shielded a massive timber door. *The crypt!* He opened the door and said, *Find your way if you can to the other end!*, and slammed the door.

It was pitch black. I sat down on something, or was it somebody? So scared, I was nearly crying. Then, after a minute or two, I began to see things. Well, no bombs down here—just dead bodies. Yes, lots of them, but in stone tombs or coffins lining the walls. The floor was large uneven flagstones that I tripped on a bit, then picked my feet up. The passage looked straight, so I kept a hand on the coffins on the right side and felt my way along, *then a light?* Noises, a star? Bethlehem? *Well it was a church!*

The other boys were down here too, but hiding and making spooky noises. I kept going, freaking and blubbering. At last, the light turned into a partially open door. I pushed on it and stepped into the warm evening light. The other boys followed me out laughing and jeering at me, but not for long for around the corner of the church came the vicar, the Reverend Carnegie, a relative of the chap who made it big in America.

"I instructed you boys that this nonsense was to cease."

He clipped the nearest two boys around the ears with religious fervor.

"This hazing will stop immediately! You boys get along home. And poor Master Short, are you all right?"

I suddenly realized he was talking to me.

"Oh. Yes, sir. I'm fine."

"Well, how about some hot chocolate in the vicarage?"

I had two delicious mugfuls in his kitchen then ran home thinking to myself: *no bombs, not too dirty, they got caught. Piece of cake and hot chocolate!* I soon became friends with all the choir boys and the two senior members who were brothers kept an eye out for all of us.

The real kicker to this is that when we moved to Mill Hill from Hampstead in 1949, I had to leave the choir. And, of course then it became world famous with recordings and performances all over Europe and the US, and Martindale Sidwell reached an exalted status as choirmaster, organist, and conductor—in high demand as a guest artist in choral music. It seems that I was the voice that was holding them back!

World-famous organist and music director who—during his long life—had a strong influence on post-war British choral music and thousands of singers young and old. After my brief, gruff beginning, he turned out to be a genteel director.

SHORT'S STORIES

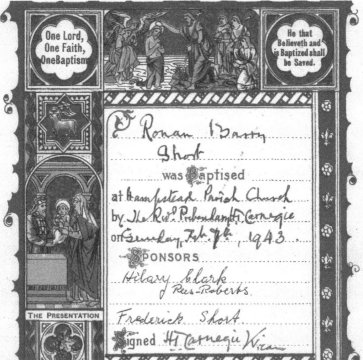

Above: Entrance of the church in the 1980s with our two daughters, Allison (left) and Renna . Below: I wasn't baptized at birth due to the war, but 3 years on at the same church where later I became a choirboy.

THE DAY THE CIRCUS
CAME TO TOWN

Two or three times a year, a travelling fair came to Hampstead Heath. It was usually on the long flat area along the Spaniards Road. Now fairs in Britain are different from those in the US. There are no animals, except mangy two-legged ones after dark, and no displays of radishes, rutabagas and rhubarb. No, English fairs are well—sleazy, dirty, *watch your back* affairs but—lots of fun. After dark even the policemen walk around on alert and in pairs. There are gypsy fortune tellers from places like Romania and Rotherham resplendent in strange hats and ornate antique caravans. There are pickpockets, youths with enormous dogs, lots of frankly dangerous rides, scamming side shows, and Texas Bill Junior's Wild West Knife and Axe throwing Act—*Roll up Roll up!*

One day, I was in Dad's office at the Institute when a large man came in wearing a tan buckskin jacket with full fringes.

"Hello," he said, "I'm Wally Shufflebottom."

I stifled a giggle.

"I'm at the fair on the heath and one of my acts has, well, we've lost a key component of our sideshow—our rats.

"Yes," he continued, "one of our daytime acts is my daughter lying in a large coffin-type box wearing a skimpy swimsuit with a dozen or so large brown rats crawling over and around her. *Oh yes!* People pay money for this and it keeps the cash coming in until the sharp-shooting and knife throwing starts in the afternoon."

Now I was paying attention!

"Well, something happened and our rats died—all of them. We made a funeral pyre and gave a circus ceremony for them, but I need more rats right now. Could I purchase a dozen or so from you Mr. Short?"

Father thought for a minute, but said, "Purchase? Unfortunately, no, mine are government rats, but maybe I could *lend* some to the Wild West Show".

A short time later Texas Bill Jr. was leaving with a cardboard box with a couple of holes and a slight squeaky noise emanating from it.

"I know they're white, but I'll fix that with brown shoe polish mixed with whitewash".

He was extremely grateful and gave us free admission to all the shows while they were in Hampstead.

A couple of days later, Dad took us three kids to the fair during the day and sure enough there was Mr. Shufflebottom, alias Texas Bill Jr., now in a white buckskin jacket with even longer fringes at his stall.

"Ah! Welcome," he said, immediately ushering us past the staring people in line with, "You know—VIPs!" And in we went.

Well, inside people were recoiling with disgust. His shapely daughter was in the coffin with her eyes closed, bored I'm sure, but the rats all had terrible-looking coats of streaked brown hair and in parts it looked wet. It turned out it was wet and it was coming off on said curvaceous daughter, who didn't seem to care at all; but the patrons were aghast.

"They're shitting all over her, how disgusting!" a large woman shouted. "This must be stopped! Where are the authorities?"

Two teenage boys were thoroughly enjoying it. By now, due to the continuing kerfuffle, the line had tripled outside. We stayed for a while then left—not a stain on our conscience! The *wet-stained* rats became a staple of the side show for many years. It is of course a custom of *carnies* everywhere to *co-opt* any and all *advantages* they come across.

Take for example Wally's father: William Shufflebottom, born in Yorkshire in about 1860. He grew up with cowboy comics and novels and later adopted a western theme for his circus act and, in fact, went to America and appeared in *Buffalo Bill's Wild West Touring Show* known as "Texas Bill." I always wondered if he spoke during the show, and whether the patrons understood a word of his broad Yorkshire accent....

"The Texans": Wally Shufflebottom, his wife Cissie, and a showgirl practicing for their Wild West Show (circa 1960).

'RIO' IN HAMPSTEAD:
COCA THE CARIOCA

Hampstead, and indeed the whole of Northern Europe in the winter in the late-1940s, was a dour, dark, wet, gloomy place. As young kids, we hardly noticed it most of the time. We rarely had special winter clothing, no rain gear, but always a sailor's oiled cotton hat: a *sou-wester* was offered. Almost everything we wore was made of wool. So, as all mariners know, when it gets wet (and it inevitably did!), you stay warm.

I even had a swimsuit made of wool until I was about eleven. Then the laughter got to be too much.

One morning on a Saturday, Mum said, "Well, that's good—we have some PTA stuff to take to Coca's."

"Oh no, do I have to go?" I pleaded.

"Yes, you'll like it. We'll all go. "Gill, get Sally's good shoes on," she insisted. "You can all do the mamba!"

"Mum, that's a deadly snake in Africa," I snorted.

"Get your jacket," she said.

It wasn't too far: down to the church, then about half a mile farther along Frognal Way. We walked in the rain; Mum had her overcoat on and a nice green hat, but she was wearing different shoes. She seemed to step lightly even in the wet.

It was just a normal-looking house, but with a bow-fronted window and the lights were all on—unusual—and two or three cars were parked outside, even more unusual. Through the brown garden gate we walked and then knocked on the plain gray door.

The door opened and *Vavoom!*—a cacophony of noise and bright colors swarmed over us as we stood gray-faced in the rain. There was Coca herself. A complete copy of Carmen Miranda but no fruit in the hair—yet. Tall, willowy, she embraced all of us; when she got to me she clutched me to her chest…she seemed to be falling out of her shiny silk dress. *Oh, God!—was I embarrassed.* Red-faced, I quickly extricated myself and stepped past her into the hallway where the loud music from the living room swarmed over me like hot sunshine!

Gillian, aged ten, and Sally—almost five—quickly started moving to the *snake* music, which turned out to be the *mamb-o*! They were soon swept out into the room by other women. It was all women looking luxurious, and they were all foreign: Spanish, Portuguese, French. They were all dancing the polyglot. And *horrors*, I was the only boy. The rain poured down outside and suddenly I was thrust onto the dance area by a short, chubby woman who had a smile from here to there.

" Bon dia," she said, "it's the Samba, Carnaval in Rio!" I understood little but soon I was moving to some sort of rhythm and *darn!*—actually enjoying it. Many years later, I would get the full force of Carnaval when, in 1969, after being released from jail in Bolivia, I finally got to Rio de Janeiro!

The dancing went on, Rose stepping nimbly in her dancing shoes, and my sisters thoroughly joining in and amused by the whole spectacle. Hours passed and everyone except us kids was drinking wine or beer.

"Oh," said a tall woman in a bright red dress, "it's still raining and it's getting dark too!"

"Dark!" said Rose. "Oh my, I've got to get my husband's dinner!"

Others acknowledged and the party slowly ended….

SHORT'S STORIES

We said good-bye and *obrigado* (whatever that meant) and stepped out into the wet grayness of North London, no streetlights on yet, and pouring down. We were soaked when we got back. We changed our clothes, putting the wets on the hot radiators, and helped Mum get dinner ready.

When Dad came home we sat down to eat.

"What did you do today?" he asked.

"Oh, nothing much," said Mum. "Just went to Coca's and delivered the PTA booklets."

"I see," said Dad. "Quiet afternoon in the rain then?"

We all sat there for a minute or so with the hot sun, the dancing colors, and infectious music all pounding in our heads, then Sally started to giggle and our vibrant time in *Brazil* all poured out to great laughter.

Especially when little Sally-lu said, "Rain? It was sunny all afternoon where I was!"

Saturday Morning Pictures and Starring in a Movie

On most Saturday mornings, all three of us would take a long walk down Fitzjohn's Avenue to the Odeon cinema at Swiss Cottage, another part of Hampstead. Once in there, we found *Uncle Edward* playing the mighty Wurlitzer organ. The words to songs would be projected onto the curtains and we would all sing our hearts out for half an hour or so until suddenly the organ started to sink below the stage and the curtains parted. As if on cue, we all started screaming. On came the usual couple of cartoons, then a serial about kids foiling crooks in the London docks, and then a *big picture*— usually a western or a story about kids in America somewhere.

We loved the Saturday morning pictures, but they left us with many questions: Why were the Indians always bad? If you had a station wagon, where was the station? Why did all the boys wear long trousers? Why did everyone live in neat tidy houses with manicured lawns? Why did they talk funny? Some of these questions have never been answered!

One Saturday, Michael Faulkner and I were at the cinema when unexpectedly an usherette moved us down to the front row. *Uh-oh*, what had we done now? There was a rather scary movie on and we were soon reacting with faces and hands. There was a figure just off to our left furiously taking flash pictures. We noticed it, but carried on having the snot scared out of us. It was Gill's best friend Judy's mother, Jill Craigie, the well-known photographer taking stills for a documentary about the effects of movies on children. Television was brand new and some believed you could get epilepsy at the cinema, or at least the flu, and cancer from a cathode-ray tube. I later saw the pictures in the *Hampstead and Highgate Express* (now just the *Ham and High*), looking really scared, but no money and no fame!

My sister Gill's best friend was named Judy. Her mom, Jill Craigie, was a very successful avant-garde photographer and film-maker who was married to the leader of the Labour party, Michael Foot, for over 50 years.

THE PEARLIES

CHARACTERS OF HAMPSTEAD I

The Pearly King and Queen of Hampstead designation has been in the Mathews family for many years. Pearly royalty is a tradition started by a road sweeper named Henry Croft. In about 1903, he started sewing pearl buttons onto his drab brown suits. It quickly caught on and in many London boroughs, cockney street traders started to cover their outer clothing with thousands of shiny pearl buttons. They were known as Pearly Kings and Queens. It was a kind of working-class reaction to the majesty and pomp of the Royal Family. Pearlies became a tradition handed down in families and still exists today.

The Pearly royalty of Hampstead in my lifetime was Bert and Emmie Mathews and family. They lived in a small Victorian cottage off Streatley Place, a footpath on our way to primary school. We would see them occasionally all dressed up, shiny and sparkling, going to events or parades. Bert's brother Ted worked at the Institute in the warehouse area: *the stores*. Sometimes if there was an event after work he would carry his elaborately decorated suit into work and wear his *pearly* hat behind the counter. The *pearlies* were a joy to behold. They were always smiling, happy and waving, I knew, directly at me.

The Pearlies were working-class Londoners who developed a tradition of wearing highly decorated clothing while collecting money for charity.

Professor C.E.M. Joad
Characters of Hampstead II

Cyril Edwin Mitchinson Joad was a Hampstead institution. He rose to fame in Britain during the war on a BBC radio show entitled *The Brains Trust*, a program with a panel of experts fielding questions sent in by listeners. He was the originator of an expression that became a catchphrase in Britain in the late forties and fifties: "It all depends what you mean by…." Also on the panel was Julian Huxley, the scientist, and a retired Royal Navy Commander Alistair Campbell, plus occasional guests. The show was broadcast live on Sunday afternoons. It was a peak hour for audiences, as there were no TV broadcasts or professional sports on Sundays.

Dad would sometimes go out on the heath with the so-called Professor Joad. Joad usually wore a brown tweed suit, always swung a cane, but never appeared to need it. Dad was very sports-minded and would often go for these walks with the Professor before or after a soccer game.

Joad was a humanist and an atheist and had *apparently* written several books on the subject. We thought he was flaky, nuts. *Why would he be seen with different ladies? A Professor? Nah! Of what?* But Dad tolerated him for the walks and *getting away from the kids—what an idea!* Mum thought he was *poor and virtually homeless*, but we later found out he had wealth, a nice house,

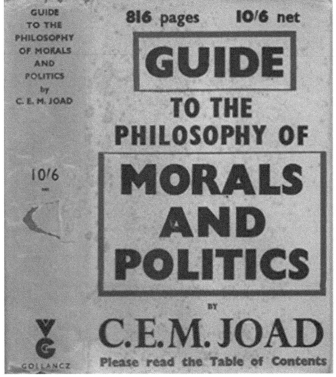

GUIDE
TO THE
PHILOSOPHY
OF MORALS
AND
POLITICS
by
C. E. M. JOAD

816 pages   10/6 net

GUIDE
TO THE
PHILOSOPHY OF
MORALS
AND
POLITICS

10/6

BY
C.E.M. JOAD
GOLLANCZ   Please read the Table of Contents

Joad's *Guide to Philosophy* was one of over 100 books, articles, and essays he wrote during his lifetime.

a wife and three kids and many mistresses. He was an ardent pacifist and rarely seen alone.

He was a voluble speaker on the Brains Trust, willing to opine on almost any topic and sound pretty good. Among him and Huxley and other guests, they discoursed on many subjects, but it was Campbell who most often would take the cake.

A question about allergies was asked once and Campbell quickly retorted: "I have one—when I eat marmalade my head steams!"

The others were quite put out over his absurd and simplistic answer, as they were readying lengthy replies to this complicated subject. Campbell was dismissed from the panel for suggesting that scientists should replace animals at the ongoing Bikini atomic tests in the South Pacific, and that's what they should be wearing.

It took a while for the reality to permeate down to us young people that Professor Joad actually was a real *Perfesser* and with high qualifications. He was the head of the department and a full professor of philosophy and psychology at Birkbeck College of London University. But even he was finally dismissed from the BBC after being convicted of travelling on a train without a ticket and fined two pounds. It all depends what you mean by...*Brains?*"

One day Professor Joad said to my father, "Where is your son going to go to school when he leaves New End. He's intelligent isn't he?"

"Oh yes," said Dad "very intelligent."

"As you know, Doug I'm well connected to University College School here in Hampstead. It's a shoo-in from there to the uni itself. I could put a word in with the governors and possibly get an interview."

"Oh wonderful, thank you," said Dad, thinking it was all nonsense.

But a few weeks later we were saluted by the gatekeeper and driving through the high, ornate gates of the school, past the immaculate playing fields, and walking into the smell of oak floors, brown furniture, and grandiose Victorian high ceilings.

We waited a few minutes during which I fidgeted with my tie (yes, a tie) and jacket and then I was shocked to hear, *Master Ronan Short!* called out by a stern-faced, uniformed attendant. Dad nudged me and I realized that was me. I jumped up and walked slowly across the polished oak into a large paneled conference room. The double doors closed behind me with an expensive click. I was then invited to stand there on a marble circle facing five men in dark suits.

SHORT'S STORIES

A number of questions were asked by what seemed to be the dullest, most formal group of men I had ever seen. I think I gave fairly good, if concise, answers to all the questions and it was obviously coming to an end when the head *inquisitor* said. "Now, tell us Short, why do you want to go to University College school?"

"Oh," I said glancing up at the dusty trophies and gray marble scrolls of achievement, "*I* don't want to come here, my *father* wants me to come here!"

Summary dismissal are the words that spring to mind. When we got home, I changed out of my stiff *interrogation* clothes and went out to kick a ball around while my parents fumed. It was a sunny evening. I played for an hour or so, scored three goals against international opponents and the brick wall, had a nice chat with Albert the whistling stoker, and went upstairs to the flat. Nobody said a word, but Mum, with a slight squeeze, brushed my shoulder gently as she placed my dinner on the table.

I guess it really does all depend on what you mean by... intelligent.

Eric Arthur Blair
Characters of Hampstead III

For more than two hundred years Hampstead has been a center of creative expression, social activism, and political foment. It contained a vast array of wealth and non-wealth! There was landed gentry of all shades and near homeless men on bicycles selling onions or repairing damaged chairs with cane or rush seats.

Among those hovering at the lower end of society was also a certain Eric Arthur Blair who lived there in the 1930s. He was a socialist, friend of Michael Foot the Labour MP, who was Judy's father, and also a strong supporter of the working classes. To get the full ambiance of homelessness, Eric would sometimes dress in tattered, dirty clothing and haunt the filthy doss houses of the down-and-outs in the East End of London. In Hampstead, he worked in a bookshop off Pond Street, wrote book reviews, and was known to visit the King of Bohemia public house, just down from the Tube Station. He would stop in there on his way back from the East End dressed in dirty ragged clothing. My father and other masons would be there also during this period, as the Freemasons Masonic Lodge met regularly upstairs. Sometimes Dad and others would have to intervene when people thought Eric really was homeless and bugging other patrons. They regularly shared a drink or two and Eric would spread his socialist views and radical opinions to anyone willing to listen.

Eric wrote a working man's history of Yorkshire, *The Road to Wigan Pier*, *Down and Out in Paris and London*, and the dystopian novels that became bestsellers around the world: *Animal Farm* and *1984*. Desiring anonymity, he wrote under a pen name derived from the King's name and a river in Essex: George Orwell. Eric Blair died young in 1950; he was 47.

Committed socialist who occasionally dressed in scruffy ragged clothes and mingled with the 'down-and-outs' in East London to gather material. He became famous shortly before his early demise.

THE MOVE TO MILL HILL

On a bright sunny spring day in 1949, the entire Institute moved to the new facility in Mill Hill. Well, our family moved that day, some scientists had already gone and many truck loads of equipment and research animals followed us. I was sent downstairs in the lift and I thought Dad was getting the car but, no, idling there in the driveway sat a long red tractor and trailer with the name *Beck and Politzer* all over it.

"Up you get, son, I'll see you in a while," said Dad, patting me on the shoulder and helping me climb into the cab. He gave a cheery wave and went back inside.

The cab had one long bench seat and at the other end was a large black man. Nervously, I closed the cab door and sat there thinking of all the things that my parents occasionally had said about foreigners: *They're from Germany, I hope they didn't bring any germs*. Or, *yiddle for diddle*, meaning they're Jewish so they'll swindle you. Or the immortal line for me personally, much later when I married Barbara Lynn Rinker: *Ronan has married an American, but she's very nice!*

"Good Morning young man, how are you today?" came forth a deep voice out of the smiling face at the wheel. "So you're going to be my navigator today, are you?"

"I don't know the way," I stammered. I had never been this close to a black person. *Were they dangerous? Did my mother know?*

"I'm Jacob and that's all right mate, I've already done about ten trips to the new Institute. I know every traffic light and manhole cover on the way."

We shook hands, mine disappearing into an envelope of power and strength. He was from Kingston, Jamaica and had served in the British Army during the war. When it ended he had stayed in England and sent for his family. He turned out to be a charming, educated man with a gift for acting and theatre. Although driving a truck was a good job and he enjoyed it, he hoped one day to break into show business.

It did not take long to drive the seven miles or so. Down the hill from Hampstead Heath to Golders Green, then across the main east-west road, the North Circular, and up through the traffic to Hendon, squeezing the rig around the Mill Hill East roundabout and going past the Underground station, the end of the Northern Line. Then it was up the steep and long Bittacy Hill, past the Army barracks to the Ridgeway. We cruised by the Adam and Eve Pub, a place that would become in itself a center of scientific and imbibing interaction, and then turned right into the Institute.

Jacob backed the truck effortlessly up to the flat and there to my surprise was one of Dad's sisters, my Auntie Marge, standing in the doorway complete with mop and bucket and a cigarette. It was always a pleasure to see her smiling and joking—and coughing! She had come to stay and help hang curtains, clean, and paint our new home. Uncle Sid was out of work again and we were assisting their finances for a week or so. The three of us unloaded our furniture and personal belongings from the truck into the various rooms and then of course had tea and chatted to Jacob. He still had to move around to the main entrance and get the remaining equipment unloaded, so after a while he swallowed a piece of Auntie Marge's cake, followed by another handshake with my small hand in his. We made our farewells and I wished him good luck. His last words were, *Go to Jamaica boy, it's warm there!*

The flat had three bedrooms, with a kitchen that had a truly wonderful unobstructed view of playing fields, farm land, and a ridge several miles away. It was in fact the original Green Belt. The flat also had separate living and dining rooms.

Auntie Marge pointed to a door and said, "That's your bedroom."

I thought she must be mistaken, but Gill and Sally it turned out would share a larger one. The whole flat had wood parquet flooring and central heating. There was also a bathtub, inside a bathroom! At the end, next to another window with a wonderful view, sat a wonderful object, a toilet bowl—with seat! We were informed later that we had to clean this one.

"Really Dad?" I asked in disbelief.

"Yes , son the stuff's under the sink. Get cracking."

The place even had a name: *The Bridge*, which wouldn't last long—but I'll get to that catastrophe.

Just six miles from Hampstead, the new Institute was used by the WRENS (Women's Royal Naval Service) during the war. It was not fully equipped for medical research until around 1949 when we moved there.

WHY WAS I NAMED RONAN?

I could say I was named after a great Celtic poet, or the saint who's buried in a village in Brittany. The truth is a little more mundane. In 1929, a bequest was made by Lord Justice Ronan KC of Dublin for sufficient monies to build an extension to the National Institute in Hampstead. This would be a center for animal husbandry and research. It was an architect-designed brick addition and attached by two walkways to the second and fourth floors of the original structure. It was named the *Ronan Building* and known as such by one and all. There was easy access from the parking lots, and the ground floor incorporated an incinerator for animal waste and large walk-in autoclaves for sterilizing cages, tools, and equipment. This addition served its purpose well, despite later being called the ugliest building in North London by leading social and political figures who resided in our neighborhood.

When I was born in 1940, Sir Charles Harington, the Institute director asked my parents if they had a name for *that beautiful child*. They said no, and he added that the Institute would eventually move to Mill Hill and who knows what would happen to the buildings in Hampstead. Would they consider a continuation of Lord Ronan of Dublin's name by bestowing it on their son? After, perhaps, thoughts of naming me Archibald

WHY WAS I NAMED RONAN?

Algernon, they quickly agreed. So I was honored, or stuck with, Ronan. For different periods of my life I've been either Ron or Ronan—the first being affectionate and the second usually an admonishment by my mother: Ronan!-Come here at once...."

As I said, our new flat in Mill Hill was named *The Bridge*. I liked the idea of living near a building that didn't bear my name. During the war, the entire Institute had been taken over by the Admiralty for the Women's Royal Naval Service, so naturally the commanding officer's quarters was *The Bridge*. My satisfaction was short-lived: soon after we were settled in, workmen appeared and replaced the sign at the bottom of the steps with the new one. Horrors! *Ronan Cottage*, it read. Sir Charles Harington's persistent devotion to that Dublin Lord had appeared again.

My name is quite common in Ireland, with many writers and sports figures being graced with it. It is relatively unusual in the U.S. and even rarer in Alaska. So that's all about *Ronan*. *Underwood* is still another matter for further discussion.

This aerial photo shows the grounds of the NIMR, including Ronan Cottage in the upper left.

I wish to thank several people for their encouragement, assistance, and patience.

Joey Becerra, my son-in-law, for sowing the seed of this memoir.

Our daughters, Renna Rose Becerra and Allison May Long, for their love and support.

My wife Barbara Rinker Short for all of the above.

My editor, Bryan Tomasovich, for his thoughtful guidance and direction.

Ronan Short was born in London, England soon after the start of World War II. After twenty-three years of an event-filled youth, he emigrated by boat from Liverpool to New York, where he worked in research as he had in England. He made his way via California to Fairbanks, Alaska by 1966. He was awarded the State of Alaska poetry prize in 1968 and completed an MFA in creative writing and theater in 1972. He worked, taught, married, and raised a family in the Golden Heart city. He and his wife of 40 years have two grown daughters and four grandchildren. Ronan has traveled extensively and has lived in Fairbanks for over 50 years. (He's still looking for the road to the airport!)